REPORT

ON THE

GEOLOGY AND AGRICULTURAL RESOURCES

OF THE

SOUTHERN DIVISION OF THE STATE.

By GEORGE H. COOK,

ASSISTANT STATE GEOLOGIST.

Printed by order of the House of Assembly.

TRENTON:
PRINTED AT THE "TRUE AMERICAN" OFFICE.
1857.

REPORT.

The geological survey of the southern division of the state has been continued, either in the field or in the laboratory, during the entire year.

The final report on the geology of the county of Cape May has been finished, and is now in the printer's hands. It is expected that it will be published in the early part of January.

In regard to this county, it may be stated, that although there is but little variety in its geological formations, and though it has no mineral wealth, it has in its soil and climate peculiar advantages for agricultural improvement. The price of land is low and a large portion of the best soil in the county is still unimproved. Considerable space has been occupied in describing the different fertilizers which are found, and in detailing the methods by which they may be made useful. The agricultural resources of Cape May, when properly developed, will add largely to the wealth of the state; and it is hoped that the geological and agricultural report, with the topographical map of the county, by Lieutenant Viele, will re der important service towards this object.

Several contributions have been made to the Cape May report, which add materially to its interest and value. Thos. Beesley, Esq., of Dennisville, has furnished a list of the larger wild animals of the county, and a very full catalogue of its birds. He has also commenced collecting birds for preservation in the state cabinet. Professor Spencer F. Baird, of the Smithsonian Institution, Washington, has furnished a catalogue of the fishes of the New Jersey shores. Samuel Ashmead, Esq., of Philadelphia, has contributed a set of beautifully preserved specimens of the marine algæ of our sea-shore, with a catalogue; also a list of flowering plants collected about Beesley's Point, together with specimens of a part of them. Dr. Maurice Beesley, of Dennisville, has written for the volume an interesting and carefully authenticated "Sketch of the early history of the county of Cape May." I take pleasure in calling attention to the valuable and generous assistance which these gentlemen have rendered to the survey.

Neither can I, in justice, omit to acknowledge the useful assistance and the hospitality which I everywhere received.

§ The survey of Monmouth county is nearly completed. A large number of marls have been analysed for the final r port, and levels have been taken for the proper construction of geological sections. Much material in relation to its agriculture has also been collected; and the principal work remaining, is to delineate the geological formations upon the map of the county, and to write the final report. These can be accomplished in a short time after the map is received. From the report of the topographical engineer, Lieutenant Viele, it will be seen that his field operations in Monmouth are completed, and the map may be expected soon.

§ In Cumberland county, all the townships have been partially examined; specimens of soils have been collected, also of marls and other fertilizers, and full notes of the agricultural practice and capabilities of the county have been made. The laboratory work is commenced and every preparation is made to carry the survey of the county forward with dispatch.

§ All the remaining counties in the southern geological division of the state, have been visited; examinations of the geological formations have been continued; collections of fossils, soils and fertilizers have been made to some extent from each of the counties. Especial attention has been given to agriculture, and it is hoped that a beginning has been made towards collecting and embodying, in available form, the successful practice of our best farmers.

§ The southern division of New Jersey, on account of some of its mineral deposits, as well as from the numerous and enormously large fossil bones and shells found in them, has long attracted the attention of geologists. Up to the year 1827, it had generally been spoken of as an alluvial formation. Some had designated it as tertiary. In that year Mr. Vanuxem and Dr. Morton, from facts which they had collected, came to the conclusion that the *marl* or *green sand* of this state was of the same geological age with the cretaceous formation of Europe. Subsequent investigations have confirmed the opinions of these gentlemen.

As the principal representative in our country of the cretaceous formation, various localities in the marl region have been much resorted to by geologists for the purpose of collecting its peculiar fossils. The inhabitants have also been in the habit of preserving some of the fossils, to give to persons who might esteem them as curiosities. In this manner the fossils have been very extensively scattered; and while some of them have found their way into cabinets of geology, the majority of the specimens, including among them many that were rare, have been lost to science. Of the specimens which have been preserved in cabinets, many are without any localities, being only labeled as coming from New Jersey.

With the discovery of the cretaceous formation in other parts of the United States and territories, renewed attention has been drawn to our own deposits of this formation, and it is felt to be a matter of much importance to know as many as possible of the fossils found here, in what subdivision of the marl stratum they are found, and their precise localities, so as to trace out, by comparison, the equivalents to these subdivisions, in other and remote places.

All the fossil shells and corals which have been collected in the course of the survey, have been placed in the hands of Professor James Hall, the distinguished palaeontologist, of New York, for description. Professor Joseph Leidy, of the University of Pennsylvania, and eminent as a comparative anatomist, has the fossil bones and teeth for description. These gentlemen have found many new species among the specimens, and of those which are known, they find many better specimens than have been seen before. Much interest is felt by them to have the collections made as complete as possible before the descriptions are published.

I have made exertions to secure all the specimens I could for the state cabinet, and where this was not possible, to borrow them for description. Many gentlemen have lent their aid in furtherance of this object. Dr. Knieskern has been indefatigable in making collections at Shark river. Valuable collections of bones and shells have been given to the state by John S. Cooke, of Tinton Falls, by Rev. G. C. Schenck of Marlboro, Thomas B. Jobes of New Egypt, M. T. Rue near Perrinesville, Henry Aregood of Kincora, J. J. Hummell of Shiloh, and others. Interesting specimens have been lent to the state by Professor O. R. Willis of Freehold, Mr. Hopper of the same place, Rev. Mr. Finch of Shrewsbury, Rev. Mr. Lockwood of Keyport, Mr. Elnathan Davis of Jericho, and others. Not having the fossils here, I am not able to give the names of all who have favored the survey, by the gift or loan of specimens, but it is intended that due acknowledgement shall be made to all in the published descriptions.

The fossils are obtained in digging the marl, and of course can only be got through the favor of those who own or work the pits. A little effort on the part of those located in the marl districts, in preserving specimens. might greatly enrich our state cabinet, and add materially to the stores of science. Every year valuable fossils which are dug out, are destroyed or given away as curiosities, to those who attach no scientific value to them; and they are, for all useful purposes, lost. I would earnestly appeal to that feeling of state pride which every true Jerseyman should possess, to enlist citizens in collecting and saving specimens to a form a cabinet, which shall properly represent the fossils of the state.

§Accompanying this report, is "A catalogue of plants growing without cultivation in the counties of Monmouth and Ocean, by P. D. Knieskern, M. D." This catalogue gives the names of seven hundred and eighty-one species, and a number of varieties, all "plants found and examined by" Dr. Knieskern, "within the last ten or twelve years." He says, "There are doubtless many plants that may still be detected within our limits, not included in this catalogue, especially in the portion farthest from the coast, which has not been examined as thoroughly as the pine barrens and the region bordering on the ocean."

It is desirable to publish in the final reports, as complete a list as possible, of plants growing in the state. By circulating this catalogue among botanists, so that our present deficiencies may be known, additions can be obtained from different parts of the state, and in this way it will be put in condition for final publication.

It is requested that this catalogue be printed in pamphlet form, separate from the report, for the use of Dr. Kneiskern and the officers of the survey.

PRESENT CONDITION OF AGRICULTURE IN NEW JERSEY.

In prosecuting the survey of Cape May county, it was found that the advancement of its agriculture had been so great since the United States census of 1850, that the statistics of the Census Report gave a very inadequate idea of the amount of its agricultural productions, at this time. By the favor of the Assessor's in the different townships, I was enabled to get accurate statistics, for the present year. The result is truly gratifying; it shows an advance in the agricultural products of about fifty per cent, since 1850, and the price of land has nearly doubled.

No statistics have been collected in the other counties; but from observations in the southern half of the State, I am well satisfied that the spirit of agricultural improvement is more active in almost any of the other counties, than in Cape May; and that the progress which is making is quite as rapid as it was between 1840 and 1850. (See tables A and B in the Appendix.)

The advancement made in New Jersey during that period, when viewed in comparison with that of the neighboring States, Connecticut, New York, Pennsylvania, and Delaware, was very flattering. In the staple crops, of wheat and Indian corn, we show an increase of more than one hundred per cent, which is more than double that shown in either of the other States, and in potatoes a gain is shown of seventy-nine per cent. against a gain of fifty-two per cent. in Delaware, and losses of twenty-two, forty-nine, and thirty-seven per cent. in the States of Connecticut, New York and Pennsylvania. Other crops show a fair advancement in the comparison, the particulars of which may be seen by a reference to the tables.

In the aggregate crops of wheat, rye, oats, Indian corn, potatoes, barley, and buckweat, if the amount for each State is taken, and averaged among the whole number of acres in each State, it gives for New Jersey, four twenty-one hundreths bushels per acre; Connecticut, two twenty-four one hundreths bushels per acre; New York, two seventy-eight one hundreths bushels per acre; Pennsylvania, two fourty-one hundreths bushels per acre; and for Delaware, three thirty-six one hundreth bushels per acre. There is also a very large balance in favor of New Jersey in products of orchards and market gardens. In the products of the dairy the balance is against us in the article of cheese. In live stock there is no great difference in the several States. It is remarkable that there is a diminution in the amount of stock kept in all the States above mentioned.

The comparison furnishes a most satisfactory vindication of our State, in its soil and productions, from the aspersions which it is so fashionable, for some of our neighbors, to cast upon it. New Jersey occupies a location for markets, unequalled by any other State in the Union. Lying between the great commercial centres of our country, New York and

Philadelphia, and having within her own borders much mechanical and manufacturing industry, a ready market for all her products is ever open; almost surrounded by navigable waters, penetrated at numerous points by rivers and creeks, and crossed by several railroads and canals, she possesses great facilities for quick and cheap transportation; so that for bulky, heavy or perishable articles she might have almost a monopoly of the markets.

Her soil, it has been well said by one of our citizens, in comparing it with that of a neighboring state "is easier tilled, equally productive, less liable to suffer from sudden changes of wet and dry, imbibes more freely the sun and dew, to favor the growth of early fruits and vegetables, and ripens them sooner for market."

With marls, limestones, and other fertilizers in great profusion within the State; with fish, crabs, and other matters, the spoils of the sea, upon her borders, and with contiguity to large cities and cheap means of transport for their waste manure and offal, New Jersey possesses unequalled resources for cheap and abundant fertilizers.

The success which attends good farming is perhaps the best evidence that can be adduced for our agricultural advantages. It has been shown from the census tables, that our product per acre, where the whole area of the State is taken into account, considerably greater than in any of the adjoining States. If the separate crops are taken, the average for corn, oats, and potatoes is higher than in the States adjacent; our wheat and rye are put down as yielding a smaller crop per acre, which, as an average, is undoubtedly correct. The premium crops of wheat, in all the counties where there are agricultural exhibitions, have been above thirty bushels an acre, in some of the counties they have been above forty bushels an acre for several years in succession, and there are instances in which crops of fifty bushels, or upwards, per acre have been harvested. The practice is much more common in this State than in others, of sowing wheat and rye in corn, or after corn or potatoes. A diminished crop of grain is the necessary consequence of thism ode of cultivation; though the produce of the field and the profits for the year are increased thereby. Wherever there is thorough cultivation the crops of grain are not inferior to those raised in other states. But it is in fruits, and in market-garden produce that our greatest advantages are found.

The returns for an acre of strawberries are from one hundred dollars upwards. The Burlingtou County Agricultural Society, in 1855, awarded the premium for the most profitably cultivated crop in the county, to one of strawberries. It yielded at the rate of twelve hundred and twenty-two dollars an acre, clear profit. Cranberry fields are known which annually yield to their owners three hundred dollars an acre. Large profits are also obtained from the cultivation of other small fruits. The state is noted for the production of apples and peaches, and fortunes have been made in their cultivation. Market-gardening pays well to those who engage in it. Early potatoes, which yield one hundred dollars an acre, are common, and those yielding three times that sum are not unknown. Equal profits can be obtained from the cultivation of sweet potatoes.

And yet with all these advantages and the examples of the large profits mentioned, to be found in all parts of the state, there are still two million acres, or nearly one half of the state uncultivated. Not entirely waste, it is true, but only yielding crops of wood, every fifteen to thirty years, which in growth may average a cord, a year, for each acre, but which on account of fires, late frosts, &c., probably produce to the owners, not half that amount.

In view of the facts presented above it is astonishing that so little attention has been paid to these unoccupied lands. The greater portion of them are not lighter or poorer than other soils from which persevering and skilful husbandry is now drawing the largest and most certain returns. And they offer the great advantages of healthy locations, and contiguity to the older settled portions of our country.

The large body of uncleard land, which occupies the central portion of southern New Jersey, is narrowed on all its borders, every year by the inroads of the farmer, and many large and productive farms are now found, where but a few years ago there was only unbroken forest.

Could the productiveness of these lands be generally known, and were they properly opened to markets, by means of roads and railroads, it would be the means of saving to the state every year, hundreds of useful citizens who now leave us to seek new homes in the west.

In regard to one of our staple crops, the potato, the Census Reports furnish some important facts. The general result of a large increase in the yearly product of New Jersey, of an increase in the product of Delaware and a diminution in that of Connecticut, New York, and Pennsylvania has been mentioned. On comparing the product of our state by counties, it appears that the increase was mostly in Monmouth, Burlington, Camden, Gloucester, Salem, and Cumberland, the counties in which marl is found, while in the remaining southern and middle counties of the state the increase is small, and in several of the northern counties there is a material diminution of the crop. (See table C in the Appendix.) So too, in the state of Delaware, the largest crop, and I presume, the largest increase of Irish potatoes, is in New Castle county, in which marl is found. In the state of New York there was an increase in only three counties, and these were the three constituting Long Island; and of these three the largest ncrease was in those, which there is some reason to believe, lie in the same geological formation, with those counties in New Jersey, in which there was the largest increase. In Connecticut there was an increase in Middlesex county, which lies in the valley of the Connecticut river, and at its mouth. In Pennsylvania there was a small increase in eight counties, located in the eastern part of the state, and mostly on the Delaware river, and its branches. Of the states in which there was no greatly increased population from immigration, New Jersey and Delaware were the only ones in which there was not a diminution of the crop of Irish potatoes between 1840 and 1850. By a reference to table C, it will be perceived that there has been a moderate increase of the crop in the sandy and light soils where marl has not been used; but the large increase is confined to the counties in which marl exists as a constituent of the soil, or is used as a manure.

The fact of this remarkable difference, in favor of our marl districts, is that which mainly concerns the practical farmer, but the cause of it cannot but be interesting to reflecting minds. Whether its special action is due to some component of the marl, or whether its usefulness is partly owing to the mingling of new earth, taken from beneath the surface, with the soil, is a question not easily answered. There is no effect produced on the appearance of the potato top. The crop, of choice varieties, is from seventy-five to one hundred and fifty bushels an acre, The potatoes are of gcod size, smooth and smoth-skinned, of superior quality for the table,* and not subject to the potato rot. The marls which are least esteemed for permanent improvement of the land, produce quite as good effects upon a single crop of potatoes as those which have the highest reputation. The Cumberland marls, which are not green sands, are deemed almost indispensible to this crop, and they produce potatoes of an excellent quality; though I think the average crop per acre is somewhat less than where green sand marl is used.

The value of this crop makes a large item in the whole agricultural product of the state. In an article in the New Jersey Farmer, for December, the potato crop of Monmouth county for the present year, is estimated at thirty-three per cent. more than that of 1850; or, in round numbers, at one million and fifty thousand bushels, and the average price as over seventy-five cents a bushel. At that price the whole crop would be worth seven hundred and eighty-seven thousand, five hundred dollars. The crop of the state may be safely estimated at four times this quantity, which in value would be three miltion, one hundred and fifty thousand dollars.

FERTILIZERS.

The abundant supplies of fertilizers found in our state

* Potatoes from the marl region will bring fifty cents a barrel more than others, in the New York market, and then have the preference.

are attracting more and more of public attention. The green sand marls are getting to be used over larger districts of country, and they are gradually finding their way to more remote markets.

There has been sent over the Freehold and Jamesburg Agriculturul Railroad, during the past year, 270,982* bushels of marl, all of which has found a market out of the marl district, and some of it out of the state. The high state of agricultural improvement along the lines of railroad, where this marl is distributed, is sufficient evidence of its value. An association called the New Jersey Fertilizer Company, has opened marl pits on the shore of Sandy Hook Bay, near Riceville, and have built a wharf for the purpose of shipping the marl of that locality. From this point marl can be very easily obtained by farmers along the shores of our own and the neighboring states. Marl is also dug extensively at White Horse, in Camden county, on the line of the Absecom Railroad, and the marl is delivered at places along the road throughout its whole length, greatly to the advantage of the country through which the road passes and to the state. The Delaware and Raritan Bay Railroad, which is now in process of construction, passes through the rich marl regions of Monmouth county, and when completed will open a large district of our state to the benefits of this valuable fertilizer, and will furnish a convenient outlet for it to the sea shore. There are other localities where the marl is situated, so that it can be readily shipped on board vessels, and numerous places situated in the interior, from which the best of marls could be cheaply procured if there were convenient means of transport.

The diminution in the cost of this article by the substitution of railroad transportation, for the conveyance by teams, will be understood when it is mentioned that formerly when marl was hauled from Squankum to Jamesburg, the cost at the latter place was thirteen cents a bushel. Now, parties interested propose if the railroad is extended from

*For these returns I am indebted to I. S. Buckalew, Esq., Superintendent of the Jamesburg and Freehold Agricultural Railroad.

Freehold to Squankum to deliver the marl in Belvidere at that price.

Inquiries are being made in reference to the price of marl and its probable amount, by enterprising persons in other states and in Europe. In regard to its price, it may be mentioned that the Squankum marl sells at Freehold for eight cents a bushel,* and that the New Jersey Fertilizer Company deliver their marl on board vessels at their wharf for seven cents a bushel. The White Horse marl is delivered on the line of the railroad at any point within ten miles of the pits, at ninety cents a ton. At pits in different parts of the marl region, the marl is sold at prices varying with the labor of excavating, without much regard to its real worth. From twenty-five to seventy-five cents a ton, includes the general range of prices.

The absolute worth of the marl to farmers, it is difficult to estimate. The region of country in which it is found has been almost made by it. Before its use the soil was exhausted, and much of the land had so lessened in value that its price was but little, if any, more than that of government lands at the west; while now, by the use of the marl, these worn out soils have been brought to more than their native fertility, and the value of the land increased from fifty to a hundred fold. In these districts, as a general fact, the marl has been obtained at little more than the cost of digging and hauling but a short distance. There are instances, however, in which large districts, of worn out land, have been entirely renovated by the use of this substance, though situated from five to fifteen miles from the marl beds, and when, if a fair allowance is made for labor, the cost per bushel could not have been less than from twelve to sixteen cents. Instances are known where it has been thought remunerative at twenty-five cents a bushel.

The chemical composition of the marls may aid in estimating their value, as fertilizers; though it must be confessed that agriculturists are by no means agreed upon the

*A bushel weighs about 100 pounds when first dug, and about 80 pounds when dry.

specific values which should be allowed for the different proximate elements found in manures.

The following are analyses of green sand marls, from different parts of the formation, and are selected as representatives of the principal varieties.*

	(1.)	(2.)	(3.)
Water	12.200	10.260	10.600
Silica	38.700	46.660	51.162
Prot-oxid of iron and alumina	27.690	24.921	22.300
Potash	4.467	6.818	4.274
Lime	....	2.865	3.478
Carbonate of lime	13.910	....	....
Magnesia	1.213	3.089	2.037
Phosphoric acid	1.140	3.599	4.540
Sulphuric acid	0.309	0.982	0.429
	99.629	99.194	98.820

These specimens are taken from the three principal marl beds; one from the first, two from the second, and three from the third. (1) is the marl which has been most largely used upon potatoes in Monmouth. (2) is the variety of marl which is most generally used in Burlington, Camden, Gloucester and Salem counties. (3) is the marl which is found in Deal, Poplar, Shark river, and Squankum, in Monmouth county.

In regard to the several constituents of the marl, it has been common to consider the phosphoric acid, and the potash, as the only substances of sufficient value to be taken into the account. Professor S. W. Johnson, of Yale College, in an article in the American Agriculturist for July, 1856, estimated the value of potash for agricultural purposes at four cents a pound, which is probably as close an approximation to its value as can be assigned, in the varying circumstances under which it is used in agriculture. Phosphoric acid he estimates at two different values, according as it is in a soluble or insoluble state. In its soluble form it is found in superphosphate of lime; in its insoluble form, in bones; to some extent, in superphosphate of lime and in guano. Soluble phosphoric acid he estimates at five

*For a description of the several beds of green sand, and their geological and geographical positions, see Geological Reports of New Jersey for 1854 and 1855.

cents a pound, and insoluble at two cents. Professor Way, Chemist of the Royal Agricultural Society of England, has estimated the worth of soluble phosphoric acid at eight and a half cents a pound, and insoluble at three cents. The prices of bones and of superphosphate are not very different in the two countries, and yet one of the results arrived at is more than fifty per cent. higher than the other. Such discordant results show the difficulties attending the subject for the authorities are, both men distinguished for care and accuracy in their investigations. If the estimate is to be made from the selling price of bone dust and superphosphate of lime, the prices assigned by Professor Way are not too high. There has been hardly time, since the first introduction of superphosphate of lime into our country, to ascertain the estimation in which it will be held by farmers. From the best information I have been able to procure, the amount of superphosphate annually manufactured is increasing. The price has diminished slightly within two or three years past, but it is still as high as the English. Bones, bone dust and bone turnings are increasing in price, and are more highly valued by the farmer every year. A reason for the differences in the value attached to phosphoric acid may probably be found, in the fact that its effects as a fertilizer, are much more perceptible upon some crops than upon others. Wheat, rye, and oats are known to be but little benefitted by it; while, on the contrary, its effects upon pastures, root crops, and garden vegetables, are entirely satisfactory. According, then, as one or other of these crops is the leading one with the farmer, will the phosphoric acid be esteemed. From the peculiar location of our state in relation to markets, and the nature of our leading products, I am led to the opinion that the prices fixed by Prof. Way are the nearest correct, for our uses. In the case of marl, however, even this is too low an estimate for the phosphoric acid. The acid is in its insoluble form; but it is disseminated through the marl in small particles, which are in a pulverulent state,—just as the superphosphate must be, as soon as it is washed into the soil,—which is a form much more available for growing vegetation, than the hard and slow-decaying fragments of bones which have been crushed in a mill.

It is a safe estimate, at least, to rate the phosphoric acid in the marls at five cents a pound, the price assigned by Prof. Johnson for the soluble acid, and in relation to the potash, I think it should be put still higher.

As the analyses may be considered to give the absolute weights of each of the different constituents in one hundred pounds of marl,—its value will be easily calculated, by multiplying the numbers in the table by twenty, and then by the price mentioned above. Thus in (1) the potash, $4.467 \times 20 = 89.34$ and $4 \times 89.34 = \$3.57$: the phosphoric acid $1.14 \times 20 = 22.8$ and $22.8 \times 5 = \$1.14$: and the sum of the two is \$4.71; the value of one ton of the marl, according to the above prices. In the same way (2) is worth \$9.05, and (3) \$7.96 a ton.

There are, however, other constituents which are of importance. Ammonia, which is acknowledged to be the most costly of all the fertilizing substances applied to soils, and which is indispensible to thrifty vegetation, is found, in small quantity, in all the marls.

The sulphuric acid, set down in the analysis, is usually combined with lime, and is in the form of sulphate of lime or plaster, a valuable fertilizer. In some of the marls there is a considerable quantity of carbonate of lime, in fine powder. This is also an excellent manure.

The oxid of iron and alumina, in their action, are not well understood. It has been asserted by some chemists that the protoxid of iron, in the marl, is changed to a peroxid by the action of air and moisture, and in the change ammonia is generated. That oxid of iron is an absorbent of ammonia is generally believed. In my report of last year I called attention to the fact that in the products of a given measure of land, a very much larger quantity of oxid of iron was taken up by potatoes than by any other commonly cultivated crop; and that this fact taken in connection with that of the large quantity of oxid of iron in marl, which is specially

2

adapted to the growth of this crop, appeared to have some significance. I hope to investigate the subject further.

Soluble silica which is liberated in large quantity, in the decomposition of these marls, has been rated very highly by some agricultural chemists.

The substances above mentioned are certainly found in large quantity, in common earths and soils, but they are not usually found in a condition to be easily made available to vegetation. The marls are soft so as to be readily penetrated by water, and though not soluble in pure water, they are readily decomposed by water containing acids, and are then soluble. Even carbonic acid, one of the weakest of acids, and which is found in water in the soil, is sufficient to decompose them, and to bring the various constituents into the condition necessary that they may act upon growing plants. It is believed that the other constituents of the marl which have not at all entered into the estimate of its value, increase its fertilizing power, both by direct action, and by their agency in absorbing and retaining other important elements of vegetable nutrition; but the precise value which is to be attached to them is not well understood.

The amount of the green sand marl which can be obtained for use is very great. It underlies the whole country in a strip which extends from Sandy Hook Bay and the Atlantic ocean, on the shore of Monmouth county, to Delaware river and bay, at Salem, in Salem county, a distance of ninety miles, and which has a breadth varying from fourteen miles at its northeastern extremity to six miles at its southwestern extremity. The area included in the strip is nine hundred square miles. On account of the marl being found in earth and not in rock, it cannot be worked under the surface like a mine; and excavations are profitably made in it only where the covering of top earth is but a few feet in thickness. In the greater portion of the area mentioned, the marl lies too deep under the surface to be profitably worked at present. But wherever streams have cut down below the general level of the country, some one of the different beds of marl is exposed, and on the sloping sides of the valleys, the deposit lying nearly level, can be worked back for a considerable distance before the covering of earth becomes so thick as to render further progress unprofitable. Precise estimates have not yet been made of the amount of surface under which marl can be profitably worked, but there are certainly many square miles, even of those varieties which are generally considered the best.

In many cases a ton or more of marl is dug from underneath each square foot of surface; but, if we allow even half of this, for the amount obtained from each square foot, a square mile will yield (13,939,200) nearly fourteen million tons; an amount sufficient to supply any probable demand for years to come.

The calcareous marl which constitutes the upper part of the second marl bed, and which has been vairously designated, as yellow limestone, yellow marl, gray marl, lime, sand, etc., is extensively developed through the length of the marl district. It consists mainly of carbonate of lime. Some portions of it are pulverulent and can be worked with a shovel, while other portions are stony and can be used for burning into lime. Its value as a fertilizer is too well known to need description here.

The tertiary marls of Cumberland county are valuable as a source of manure for the district of country in which they lie. They are found on the western borders of the county and are principally dug in the valleys of several streams, which, when united form Stoe creek. The whole of the workings are comprised within a strip which extends about four miles in a northeast and southwest direction, and which is perhaps a half mile wide.

These marls are not green sand. Some of them have a large per centage of shells which are in a broken and decayed condition; others are entirely destitute of shells or of carbonate of lime. The two analyses which follow, are sufficient to exhibit their composition.

TABLE.

Silica and quartz sand, -	79.160	83.328
Peroxid of iron, - -	3.562	4.770
Alumina, - - - - -	.442	5.412
Lime, - - - - -	7.500	0.211
Magnesia, - - - -	0.884	0.668
Potash, - - - - -	1.227	0.829
Phosphoric acid, - -	0.420	0.854
Sulphuric acid, - - -	0.166	0.283
Carbonic acid, - - -	4.030	
Nitric acid, - - -		a trace
Ammonia, - - - -		0.067
Organic matter, - - -		1.967
Water, - - - - -	2.400	2.193
	99.791	100.582

The first of these marls is like most of those containing shells, in some however fifty per cent. Some have attributed their value to the carbonate of lime in the shells; but those who use them do not consider this to be the cause of their fertilizing action, and make no difference in the price of those which contain shells and those in which there are none. There is a tract of country about Shiloh, upon which lime has never been observed to produce any beneficial effect; but which has been brought to a high degree of fertility by the use of these marls. They are carted to distances of five miles, and are thought to pay well for the expense of transportation, and the first cost, which is from fifty to seventy-five cents a wagon load.

The second analysis is that of a rather remarkable substance. It is called marl by the inhabitants, and is dug in the same tract of country with the other. To an ordinary observer it would seem to be a yellow loam or subsoil, and there is nothing by which a closer inspection would be able to distinguish any difference. It is extensively used as a manure, and is particularly valued for spreading upon grass lands. Farmers purchase it at the pits for twenty-five cents a load, and haul it four or five miles for use, and think it well repays its cost.

The king-crabs or horse-feet, were mentioned in the Annual Report of last year, as abounding on some of our shores; and it was also mentioned that an establishment for making a concentrated manure, from them, had been erected at Goshen, in Cape May county, by Messrs. Ingham & Beesley. Several hundred tons of this substance were made last year and sold under the name of *Cancerine.* It is a powerful fertilizer, and in its composition as well as in its effects, has considearble resemblance to guano. The fertilizing properties of guano are generally conceded to be due to the amounts of ammonia and phosphoric acid which it contains. The per centages of ammonia and phosphoric acid contained in guano and in the cancerine, are here given.

	Ammonia.	Phosphuric Acid.
1. Peruvian Guano, - -	15.00	14.75
2. Peruvian Guano, - -	14.79	10.15
3. Cancerine, - - -	10.75	2·71
4. Cancerine, - - -	9.92	4.05

No. 1. is from an analysis of No. 1 Peruvian Guano, by Prof. S. W. Johnson, of Yale College; published in the American Agriculturist, for December, 1856.

No. 2. is from an analysis made in my laboratory. The specimen was obtained from a respectable dealer, in New Brunswick.

No. 3. is the result of an analysis of what I thought to be an average sample of Cancerine.

No. 4. is the average of three analyses of Cancerine, made here at different times.

In the article of Professor Johnson, which was just referred to, he estimates the value of ammonia at sixteeen cents a pound, and phosphoric acid in guano, at two cents a pound; this would give the prices of

1. Guano, - - - -	$ 53.90	per ton.
2. Guano, - - - -	51.39	"
3. Cancerine, - - -	35.48	"
4. Cancerine, - - -	31.16	"

In this estimate phosphoric acid is set down at a lower price than I have assigned to it in calculating the value of green sand, but as the two prices bring the guano up to its selling price, and the only object is a comparison, this variation will not materially influence the relative values of the two.

There is a difference in the state in which the ammonia exists in the two substances. In guano much of it is ready formed, and is easily dissolved by water, or volatilized by heat; while in cancerine, though the elements are present, the ammonia is not yet formed, and it is not soluble nor volatile; and so not liable to deteriorate in strength, until the process of decay commences in it. But for the same reasons the guano is quicker in its action.

The result of trials with it, during the year, have fully sustained its value as determined by the analysis. The wheat crop of southern New Jersey, last year, was from five to ten bushels an acre less than common, probably on account of the severe winter. The effects of neither guano nor cancerine were very decided upon this crop. Some very good crops of wheat were harvested where cancerine was the only manure used. Upon summer crops, Indian corn, potatoes, &c., it has succeeded in all cases.

From the results obtained, I consider this pioneer establishment to have demonstrated that we have both the material and the means for manufacturing a concentrated manure which in its quality and price is a substitute for Peruvian guano. Several thousand tons of the cancerine could be produced in a year.

The fish, particularly the mossbonkers which abound near our shores, have been much used as a manure. Applied in the raw state, they are powerful fertilizers. An examination and analysis of the mossbonker has been made here, the present fall, for the purpose of ascertaining the value of this fish, when dried, for manure.

ANALYSIS OF THE FRESH FISH.

Water, - - - -	77.17 per cent.
Oil, - - - -	3.90 "
Dry substance, - - -	19.93 "
	100.00

Analysis of the dry Fish.

Lime, - - - -	8.670
Magnesia, - - -	.670
Potash, - - - -	1.545
Soda, - - - -	1.019
Phosphoric acid, - - -	7.784
Chlorine, - - -	.678
Silica, - - - -	1.333
Organic matter, - - -	78.301
	100.000
Ammonia, - - -	9.282

There were five fishes in the lot examined, and they weighed four pounds, four and one-eighth ounces, which is a little more than the average weight of such fish during the spring and summer. They were taken in the latter part of October, and were quite fat.* The yield of oil from them is probably above that of other seasons of the year. The amounts of ammonia and phosphoric acid contained in the dry fish are sufficient to make it a valuable concentrated manure.

The amount of material for such manure which could be obtained upon our shores, has not been accurately estimated,

* The specimens were procured for me by Charles Sears, Esq., of Riceville, Monmouth county.

but it is enormous. A friend who has been at some pains in making inquiries upon the subject, estimates the amount which could be obtained at a single point on the shore, during one season, at one hundred thousand barrels. Sixty wagon loads were taken at a single haul, on the shore of Raritan Bay, the last summer. In different seasons the prices vary, between five and eight cents a bushel. The price on the shore of Monmouth county has not been below eight cents a bushel during the past season.

It would be a public benefit to have these supplies of fertilizing materials, which fairly bring themselves to our shores, put in a form to be more extensively used; and the oil and the fish manure offer fair opportunities for profit to a well conducted manufactory of these articles.

The following analyses of green sand marl were made from specimens taken from different parts of the marl formation for the purpose of comparing the composition of the green mineral. (1 and 2) are from the first marl bed, (1) from Mannington township, Salem county; (2) from Middletown, Monmouth county; (3 and 4) are from the second marl bed, (3) from Mannington, Salem county, and (4) from Shrewsbury, Monmouth, (5 and 6) are from the third or upper marl bed, (5) from Gloucester township, Camden county, and (6) from Ocean township, Monmouth county.

Analyses of Green Sand as taken from the Marl Beds.

	(1.)	(2.)	(3.)	(4.)	(5.)	(6.)
Prot-oxid of iron	8.320	16.816	21.312	21.204	14.930	17.388
Alumina	6.096	6.658	7.992	8.220	6.000	6.048
Lime	2.364	12.524	1.000	0.496	1.985	4 066
Magnesia	0 423	2.544	2.034	2.024	1.615	2.963
Potash	2.464	4.830	7.064	7.088	4.296	3.680
Soluble silica	20.200	31.200	45.912	46.936	33.960	35.356
Insoluble silica	49.920	5.640	4.008	4.220	23.680	9.328
Sulphuric acid	0.872	0.632	0.408	.411	0.878	3.124
Phosphoric acid	1.392	1.084	1.272	0.192	2.640	6.864
Carbonic acid	0.200	9.320	.000	.000	.000	.000
Nitrogen	0.047	0.043	Trace.	0.017	0.038	0.024
Water	7.080	8.920	8.090	8.500	10.640	10.210
	99.378	100.261	99.092	99.418	100.662	99.051

The per centage of matter soluble in water was ascertained to be as follows:

	(1.)	(2.)	(3.)	(4.)	(5.)	(6.)
Alumina	0.105	0.135	0.075	0.065	0.075	0.055
Lime	0.546	0.455	0.349	0.186	0.382	1.226
Magnesia	0.097	0.079	0.093	0.093	0.079	0.044
Potash	0.238	0.278	0.186	0.362	0.134	0.097
Sulphuric acid	0.876	0.415	0.398	0.406	0.813	2.876
Prot-oxid of iron	Trace.	Trace.	0.031	0.031	0.103	0.201
Phosphoric acid	....	....	....	....	0.355	0.194
	1.862	1.362	1.132	1.143	1.941	4.693

It should be observed in relation to these specimens of green sand, that they were not selected for the purpose of getting specimens of the average value as fertilizers, numbers 3 and 4 are much below the average, and number 6 is above, containing a larger per centage of phosphoric acid than any othhr specimen that I have examined.

The following analyses are of clean grains of green sand. The specimens were prepared by first washing out all clay and muddy substances that could be kept suspended in water; then drying the remaining matter at about a summer heat; and afterwards carefully picking out the grains of green sand from the particles of quartz, phosphate of lime and other substances with which they were mixed. After all the trouble taken, however, it will be perceived that there was a small quantity of sand and phosphate of lime left with the grains. The phosphate of lime is evidently not a constituent of the grains of green sand; it can be plainly distinguished from them in the mass, by the color of its particles, which are of a very light green or greenish white; but some of it adheres so closely to the green sand grains, as not to be separated by washing.

Of the three specimens here given; (1) is from the first marl bed, and was obtained at Cream Ridge, Monmouth county: (2) is from the second marl bed, and was procured at White Horse, Camden county; (3) is from the third marl bed, and was taken from a pile of Squankum marl, at Freehold

Analyses of Grains of Green Sand.

	(1.)	(2.)	(3.)
Soluble silica	45.510	50.070	41.729
Prot-oxid of iron	21.134	21.120	16.627
Alumina	7.960	7.323	5.929
Magnesia	2.480	2.866	2.958
Potash	6.748	7.370	6.066
Lime	3.842	0.312	8.026
Phosphoric acid	0.993	0.628	7.356
Sulphuric acid	1.129	0.430?	1.005
Carbonic acid	0.563	0.000	1.383
Insoluble silica (sand,)	0.850	0.402	0.909
Water	9.110	9.474	7.688
	100.299	99.980	99.656

An examination of the preceding analyses will show that the soluble silica, prot-oxid of iron, alumina, magnesia, potash and water, are nearly the same, in quantity, in all the specimens, while the other constituents are extremely variable. It seems a legitimate conclusion from this examination, that the grains of green sand are made up of the constituents mentioned above, as being constant; and that the remainder compose the material which is found mixed in with the grains, and, in some cases, adhering to them. If we assume this conclusion to be correct and sum up, in each column, the six substances mentioned, as constituting the green sand, we may by an easy calculation ascertain the amount of each of these substances in one hundred parts of the pure green sand. The following are the results of such a calculation upon each of the three analyses given last.

Composition of Green Sand, as Calculated from the preceding Analyses.

	(1.)	(2.)	(3.)
Silica	48.977	50.923	51.532
Prot-oxid of iron	22.744	21.504	20.533
Alumina	8.566	7.503	7.322
Magnesia	2.647	2.918	3.628
Potash	7.262	7.505	7.491
Water	9.804	9.647	9.494
	100.000	100.000	100.000

The close resemblance shown between the three specimens, when compared in this way, gives satisfactory evidence that the green sand is a definite chemical compound. The inference, too, seems a fair one, that the variation which is observed in the fertilizing properties of the marl or green sand, is due to the foreign matter mixed in with the grains, and which varies with the localities.

A re-examination of the white clay of South Amboy shows it to contain zirconia. Qulitative analyses of the white clays of Woodbridge and Trenton detect the same substance in them also, and it is probably a constituent of all the clays of this formation. There is reason to believe that the white clay at Trenton is produced by the decomposition of the gneiss rock, which occurs there. Crystals of zircou are not uncommon in the rock at that place.

Analysis of a fire-clay from Whitehead's clay-bank, on Burt's Creek, near South Amboy, Middlesex county.

Soluble Silica, - - -	43.937
Insoluble Silica, - -	1.703
Alumina, - - - -	38.008
Zirconia, - - -	1.403
Peroxid of iron, - -	0.747
Magnesia, - - -	0.077
Lime, - - - -	0.413
Potash, - - -	0.368
Water, - - -	13.863
	100.519

In closing the report it should be remarked that those scientific details which must give to the survey its accuracy and value, have been mostly omitted, not from under-estimating their importance, but because it is judged they will find their more appropriate place in the final reports.

The practical and economical relations of the survey are the main objects for which it was instituted, and to these this report has been chiefly confined. If I have succeeded in exhibiting the important relations they hold to the material wealth of the state, I shall have done no more than justice to the enlightened policy which originated and has sustained the work.

GEO. H. COOK,
Asst. State Geologist.

[A.]

Table showing the areas of the states of Connecticut, New York, New Jersey, Pennsylvania and Delaware, in 1850; also, the Live Stock and Principal Agricultural Productions of those states in 1840 and 1850.

	Connecticut.		New York.		New Jersey.		Pennsylvania.		Delaware.	
Areas of the States in acres	3,001.360		30.080.000		4 656 370		29,040,000		1,356.800	
Acres of improved land	1,768,173		12 408 964		1 767 991		8.623,619		580.862	
Acres of land, in farms, unimproved	615,701		6,710.120		984 956		6,294.273		375,282	
Per centage of whole area in farms	79		63		59		51		70	
Value of farms per acre	$30 50		$29 00		$43 67		$27 33		$19 75	
	1840.	1850.	1840.	1850.	1840.	1850.	1840.	1850.	1840.	1850.
Horses and mules .... Number.	34.650	26 928	474 543	447,977	70,502	63,044	365,129	352 657	14 421	14.643
Neat cattle	238.650	212 675	1,911,244	1.877.639	220.202	211 265	1,172,665	1.153 946	53.883	53 211
Sheep	413 462	174 181	5.118,777	3.453,241	219 285	16[illegible].488	1,767,620	1,822.357	39.247	27.503
Swine	131.961	76.472	1.900,065	1.018,252	261,433	25[illegible],[illegible]70	1,503,964	1.040.366	74 228	56 261
Wheat .... Bushels.	87,009	41.762	12.286 418	13,121.498	774 203	1,60[illegible].190	13.213.077	15,367.619	315 165	482 511
Rye	737.424	600,893	2,979 323	4 148.182	1.665.820	1.25[illegible] 578	6,613.873	4 805,160	33.546	8,066
Oats	1.453.262	1,258.738	20,675.847	26,552,814	3.083 524	3.37[illegible] 063	20.641.819	21,538.156	927.405	604 518
Indian corn	1.500,441	1 935 043	10,972 286	17.858,400	4,361 975	8.75[illegible].7[illegible]4	14,240.022	19.835 214	2,099.359	3,145 542
Irish and sweet potatoes	3,414,238	2,689 805	30.123.614	15.403 997	2,072.069	3,715 251	9.535.663	6,032,904	200.712	305,985
Barley	33,759	19,099	2 520.068	3.585.059	12.501	[illegible].492	209,893	165.584	5,260	56
Buckwheat	303.043	229.297	2 287.885	3,183 955	856.117	87[illegible] 9[illegible]	2.113,742	2.193,692	11.299	8.615
Hay .... Tons.	426,704	516 131	3.127,047	3 728,797	334 861	43[illegible].950	1,311.643	1.842.970	22.483	30,159
Value of orchard products .... Dollars.	$296,232	$175.118	$1,701,935	$1,761.950	$464,006	$60[illegible] 268	$618,179	$723 389	$28,211	$46.574
Value of market garden produce	....	196 874	....	912.047	....	475 242	....	688 714	....	12,714
Tobacco .... Pounds.	....	1.267.624	....	83.189	....	310	....	912,651	....	....
Butter	....	6,498,119	....	79,766 094	....	9,48[illegible] 210	....	39.878.418	....	1,055.308
Cheese	....	5,363,277	....	49,741,413	....	365,756	....	2,505,034	....	3,178

[B.]

Table showing the per centage of *Loss* or *Gain*, in several staple agricultural products, in the states of Connecticut, New York, New Jersey, Pennsylvania and Delaware, from 1840 to 1850; also, the *Product* per acre in each state, when its whole crop of 1850 is divided among the whole number of acres in the state. (The product of wheat, rye, oats, Indian corn, potatoes, barley and buckwheat, is given in bushels; that of hay in tons; and of orchard products in dollars.)

	Conn.			N. York.			N. Jersey.			Pennsyl'a.			Delaware.		
	Loss.	Gain.	Product.	Loss.	Gain.	Product.	Loss.	Gain.	Product.	Loss.	Gain.	Product.	Loss.	Gain.	Product.
Wheat	52	...	.014	...	7	.436	...	107	.344	...	16	.529	...	53	.356
Rye	19	...	.200	...	39	.137	24	....	.259	30	...	.165	76	...	.006
Oats	13	...	.402	...	29	.882	...	9	.725	...	4	.742	35	...	.446
Indian corn	...	29	.644	...	63	.593	...	101	1.881	...	39	.683	...	49	2 318
Potatoes	22	...	.896	49	...	.512	...	79	.802	37	...	.208	...	52	.225
Barley	44	...	.006	...	42	.118	48	....	.001	21	...	006	98	...	......
Buckwheat	25	...	.076	...	39	.106	...	3	.189	...	4	.075	24	...	.006
			2.238			2.784			4.211			2.408			3.357
Hay	...	21	.170	...	19	.124	...	30	.093	...	40	.063	...	34	.022
Orchard products	41	...	.053	...	4	.958	...	31	.130	...	17	.025	...	65	.034

[C.]

Table showing the number of bushels of potatoes raised in the several counties of New Jersey and Delaware, in 1840 and 1850.

NEW JERSEY.

COUNTIES.	1840.	1850.
Sussex	201.090	110,020
Warren	142 662	92 278
Passaic	78 886	79,169
Bergen	127,043	166,368
Hunterdon	121,569	78,734
Somerset	76.854	63 573
Morris	219,996	135,518
Essex	178,193	159.282
Hudson	14.478	32 885
Mercer	57,581	96.322
Middlesex	86.965	127.024
Cape May	14.394	18 548
Cumberland	31 851	137.313
Atlantic	15,932	21,645
Ocean, (formed from Monmouth, in 1850,)	....	40.371
Monmouth	273,280	813.849
Burlington	193,126	412.143
Camden, (formed from Gloucester, in 1844,)	....	373 080
Gloucester	167,525	508 834
Salem	70,644	248.315

DELAWARE.

COUNTIES.	1840.	1850.
New Castle	84,166	125 924
Kent	68.357	89 225
Sussex	48 189	90 806

www.ingramcontent.com/pod-product-compliance
Lightning Source LLC
LaVergne TN
LVHW011126110826
845150LV00008B/2259

9781418195380